D0944530

GUIDE TO FLORIDA POISONOUS SNAKES

BY
ROBERT ANDERSON

ISBN 0-932855-04-0

Library of Congress Catalog Card Number 83-70831

Photography and artwork by the author unless credited otherwise.

TABLE OF CONTENTS

ILLUSTRATIONS
(Black and White)

TABLE OF CONTENTS
(Continued)

ILLUSTRATIONS
(Color)

SPECIES
(text)

ACKNOWLEDGEMENT

To the Crandon Park Zoo and staff; Dr. Gordon L. Hubbell, D.V.M., Director – Dr. Ronald N. Sampsell, D.V.M. Veterinarian – Phillip Allen, Educational Officer. To William E. Haast, Director of the Miami Serpentarium and Venom Institute and to Dr. Gary W. Schmelz, Director of the Big Cypress Nature Center. Thank you for your valuable time, knowledge, and for the use of living specimens of the species for photographic purposes.

THE AUTHOR

An outdoor writer, naturalist and photographer for more than 40 years. Almost 27 of these years have been spent within the boundaries of Florida writing about and photographing wildlife; from the seashores of both coastlines to the fields, forests, swamplands and hammocks. His study of reptiles, particularly poisonous snakes, was kicked-off when he received a vicious bite from a healthy "and excited" 42-inch copperhead snake. This was accomplished during his first attempt to extract its venom. His interest in poisonous snakes was further encouraged through correspondence with the late Raymond L. Ditmars and Clifford Pope. He has hunted snakes with such experts and friends as Marlin Perkins, Emil Rokosky and the late Moody Lentz and Burt Tschambers. However, no personal hunting experiences are entered into this guide. Only personal observation is recorded regarding the habits and habitats of Florida's 11 poisonous species, all of which the author has had the pleasure of meeting on their own grounds. There is probably more information, contained within this guide, regarding these reptiles, than any other non-technical publication released to this date.

INTRODUCTION

In preparing this guide book my aim has been to produce a work sufficiently free from technicalities to appeal to the general reader and at the same time to include such scientific information relative to Florida's poisonous snakes as would be desired by one beginning their study as well as for those who are eager to advance their knowledge of the subject.

Many popular errors will be corrected, while others will be disposed of by a simple statement of facts, from which the readers are expected to draw their own conclusions. There will be no attempt, however, to discuss and controvert the many exaggerated stories and beliefs passed on through many generations concerning the subjects of what is believed to be the most revered forms of animal life in the Sunshine State.

If a snake is caught, killed, or seen, and any questions raised about its poisonous or harmless nature, it will be found that the presumption of guilt is against it, and that absolute proof by a highly educated person, on the subject, will be required before anyone is willing to believe its innocence. An expert insisting that the snake in question belongs to a species wholly devoid of venom would probably be met with a statement that a serious case of poisoning had been observed by the person as a result of this particular snake's bite.

Nothing is easier than misidentification of snakes. Consequently the bite might have been caused by a different kind of snake than the one discussed. If this is the case, the expert would not be in a position to contradict the accuracy of the statement. However, the expert may recall to the challenger, a number of similarly, serious cases resulting from the bite of other animals such as rats, dogs, cats, cattle, horses, and even humans which have resulted in severe swelling and inflamation, perhaps, even death.

It is clear then that we cannot always conclude a snake belongs to a venomous group from the fact that its bite results in symptoms of poisoning. Modern science shows us that such results in other individuals are due to the presence of minute organisms in the saliva, known as bacteria. The general public knows these cases as blood poisoning. The professional man refers to them as cases of septicemia. The fact is that the venom of truly poisonous snakes is only modified saliva. However, this should not lead anyone to suppose that snake venom and bacteria infected saliva have anything in common in their nature.

When speaking of venomous snakes I shall only refer to such snakes as are provided with a specific poison and an apparatus especially adapted for the introduction of this poison into the wound of the victim. Only those of our snakes are referable to this category which are possessed of a pair of movable or constantly erect, rigid fangs at the anterior or posterior end of the upper jaw.

All counties throughout the State of Florida have at least one representative of the poisonous snake groups.

FAMILY: CROTALIDAE
PIT VIPERS

The representatives of this snake family include the rattlesnakes, copperheads, and cottonmouths. These snakes have acquired the popular name "pit viper" - owing to the peculiar development seen with all the species. This consists of a deep pit on each side of the head between the nostril and eye. The pit is lined with a delicate epidermis, and connects with a well developed nerve extending backwards to the brain. These facial pits are specialized sense organs which react to temperature differences as slight as a fraction of a degree. This enables these snakes to detect warm-blooded animals on which they prey. Because of this development pit vipers are able to locate their prey and to strike accurately in the blackest of night. They have been observed hunting during the late evening testing the temperature at the entrances to burrows until one is found giving off the telltale heat from a warm-blooded inhabitant. They have also been known to seek out sleeping warm-blooded prey in tall grasses during the night-time hours. Most of the pit vipers release their prey immediately after striking, then track the animal down by their sensitive pit or by their keen sense of smell. However, when a bird is bitten these snakes will hold the victim in their mouth until the venom kills it. This is done because a stricken bird, if released, is capable of some flight and therefore leaves no trail to be followed by the reptiles.

Pit vipers have vertical or elliptical pupils like those of a cat during the daytime but during the late evening hours and night-time they become almost round.

The members of this family can be recognized by the flat triangular head, very distinct from the neck. The top of the head, with the majority of the species is covered with small, granular scales. Some show regularly, arranged head shields. All are thick bodied for their length.

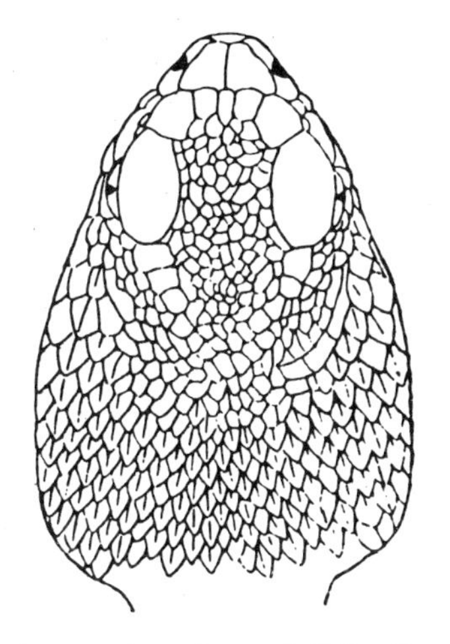

Head triangular and distinctive from neck.

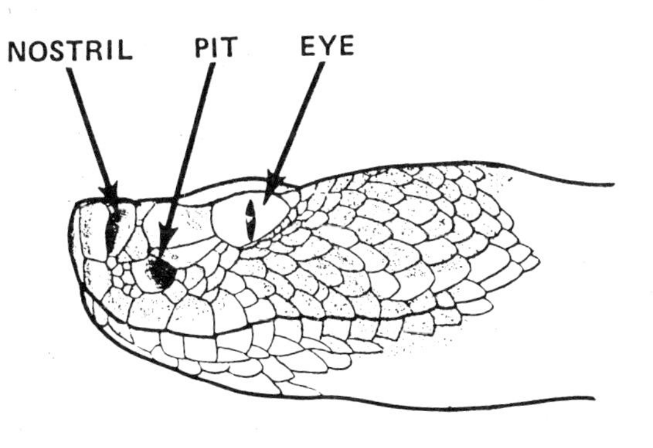

The poison apparatus of the pit viper consists of two long and hollow teeth – fangs – provided with an elongated orifice at their tips for the ejection of the venom. These fangs are the exact reproduction, in hard enamel, of the hypodermic needle. The fangs are rigidly fastened to a movable bone of the upper jaw and each contacts with a venom gland.

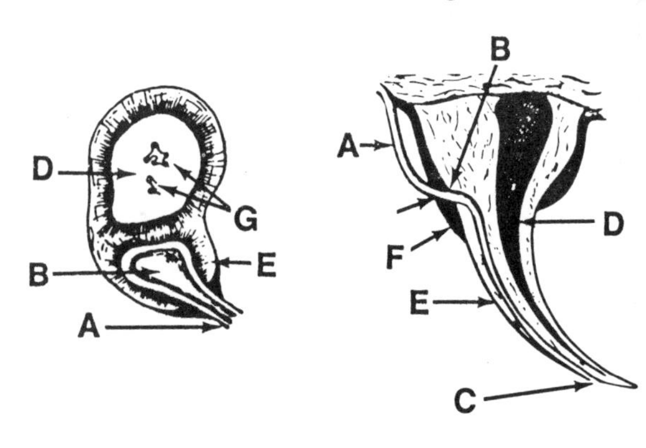

TRANSVERSE SECTION OF FANG (LEFT); LONGITUDINAL SECTION OF FANG (RIGHT). A - poison duct entering the fang at B; C - opening of poison canal near tip of fang; D - pulp cavity; E - dentine; F - connective tissue; G - nerve center.

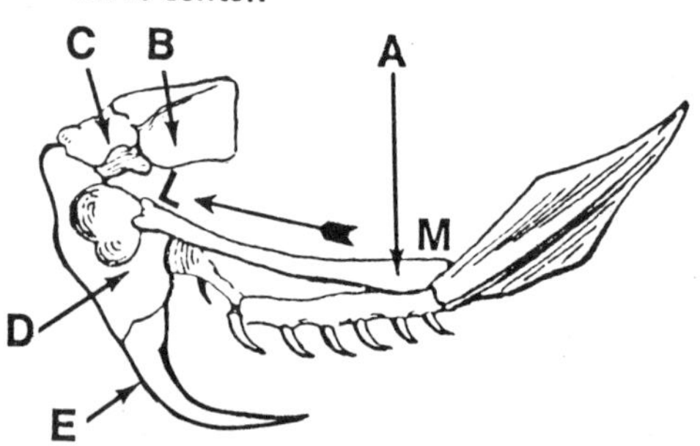

BONES CONCERNED IN RAISING THE FANG A - pterygoid bone L - M arrow marking its line of motion; B - frontal bone; C - lachrymal bone; D - maxillary bone; E - fang

The typical venom gland is found in its fullest development among the pit vipers, and is located on each side of the head below and behind the eyes. The shape is that of a flattened almond, the pointed end towards the front and below the eye, tapering to a narrow duct, which carries the poison to the inlet at the base of the fang. This duct, in its normal position, makes a sudden upward curve under the eye, descending from which it follows the posterior wall of the pit and finally passes over the rounded outer front edge of the maxillary bone, at the base of which meets the upper opening, or inlet, of the canal through the fang. The relative size of the organ may best be understood by the following illustrations. Because of

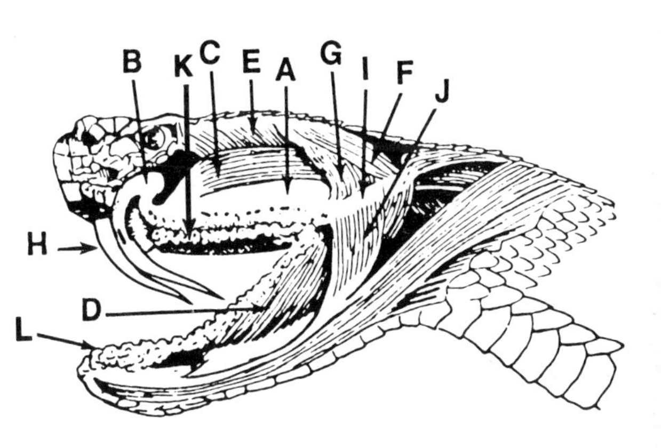

POISON APPARATUS OF RATTLESNAKE: VENOM GLAND AND MUSCLES. LATERAL VIEW. A - venom gland; B - venom duct; C - anterial temporal muscle; D - mandibular portion of same; E - posterior temporal muscle; F - digastricus muscle; G - middle temporal muscle; H - sheath of fang; I - posterior ligament of gland; J - external pterygoid muscle; K - maxillary salivary gland; L - mandibulary salivary gland.

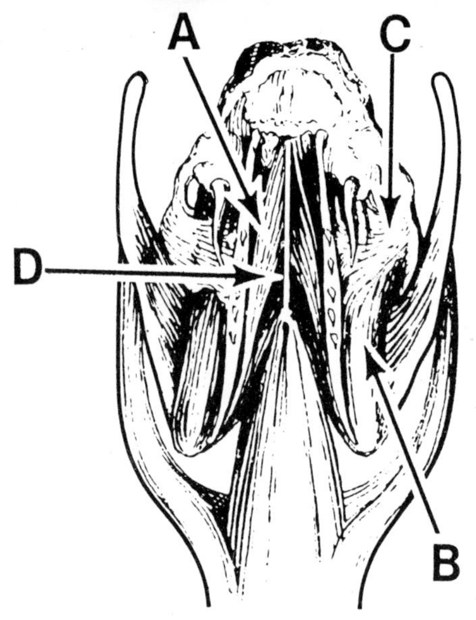

MUSCLES OF POISON APPARATUS OF RATTLESNAKE, PALATIAL VIEW. A - spheno-pterygoid muscle; B - external pterygoid muscle; C - fascial sheath of this muscle attached to the capsule of the gland; D - median ridge of base of skull.

the placement of the fangs on the movable part of the maxillary bone the fangs fold back against the roof of the mouth when the reptile's jaws are closed. As the jaws are opened, the fangs are swung forward and ready for action. The forcible ejection of venom from the fangs is caused by the contraction against the glands, of muscles which close the jaws. The ejection of venom is voluntary, and unless the reptile so desires there is no necessity in closing the jaws, to contract these muscles sufficiently to force venom from the glands. The fangs are covered with a sheath of thin, white, membraneous flesh. This is never withdrawn from them except during the act of biting.

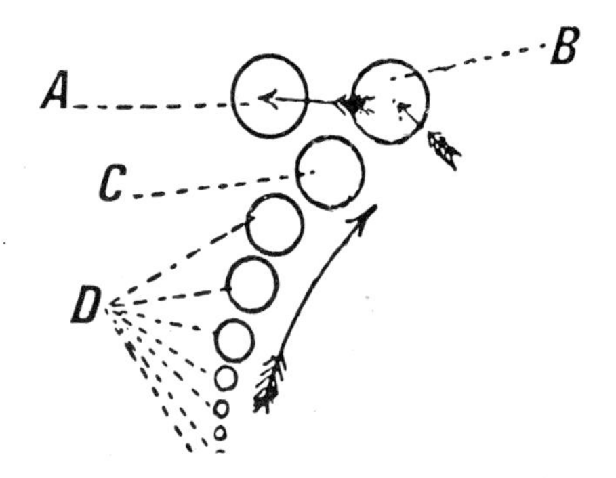

DIAGRAM ILLUSTRATING THE SUCCESSION OF THE FANGS. A - functionary fang; B - its successor; C - the next fang in order of age; D - remaining reserves.

The fangs are shed at intervals about three months apart, and by a neat provision of nature two new fangs grow into place beside the ones about to be shed and become connected with the poison glands, before the old fangs become loosened. Many specimens have been examined possessing two perfectly developed fangs on one side of the jaw, as illustrated. The old fang is usually shed by being left imbedded in the body of the prey that is bitten by the snake and is consequently swallowed with the prey. So hard is the composition that, although the bones, claws, and eventually the teeth of the engulfed animal are entirely dissolved, the swallowed fang is unaffected by the action of the snake's gastric juices. The cross - sectioned illustration shows the placement and the growth of the auxiliary or reserved fangs behind the active pair. This constant renewing of the fangs disposes the common supposition that a snake can be rendered harmless by removing its fangs. Though the main part of the fangs may be removed the snake is not rendered even temporarily harmless, for the poison is discharged in the act of biting from the base of the extracted fangs and the teeth of the upper jaw, normally employed by the snake in swallowing its prey, would produce laceration through which the venom could come in contact with the blood.

As previously explained, the fangs themselves are not movable, but are rigidly attached to movable bone. In the act of striking, the jaws are opened to such an extent and the fangs are so elevated that their tips almost point directly forward. In striking toward a perpendicular surface the snake literally stabs with the fangs and instantly draws back to a position of defense. If striking toward a rounded surface or a small object, the jaws close upon it enough to imbed the fangs, but so lightning-like is this movement that the opening and closing of the jaws can barely be followed by the human eye. The mouth is never opened until the head starts forward, and it is during the latter part of the strike that the jaws are thrown open to the extent previously described.

At most the pit vipers strike about one-half their length when delivering an accurately aimed bite, and generally strikes a much shorter distance in proportion to its length. If, however, they are tormented to a state of frenzy a strike could equal two-thirds their length, but such deliveries are usually wild and done in an aimless manner. The amount of venom ejected varies with size and species. The poison glands of all pit vipers are believed to contain enough venom to kill a man but it is extremely unusual for any of these snakes to empty these glands during a single ejection. Some may expel only a fraction of a drop. This, however, is enough to cause pain and discomfort to the person bitten.

The venom of pit vipers are of a hemotoxic action. A toxin capable of destroying blood cells in both warm-blooded and cold-blooded animals.

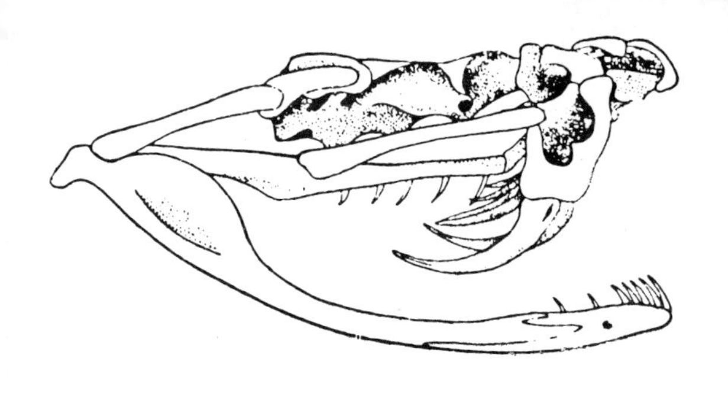

SKULL OF RATTLESNAKE, SIDE VIEW.

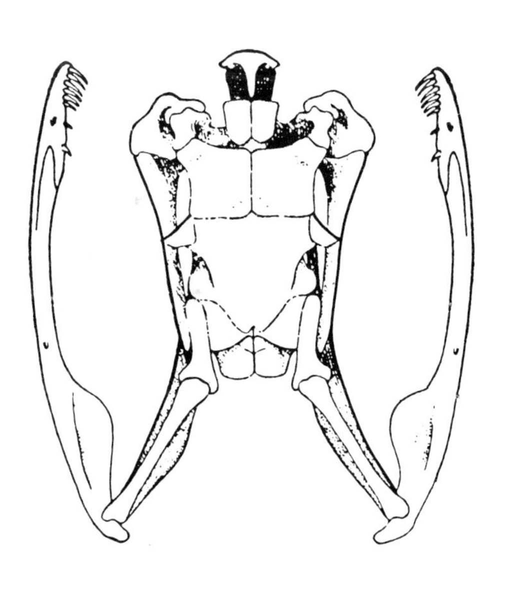

SKULL OF RATTLESNAKE, TOP VIEW.

EASTERN DIAMONDBACK RATTLESNAKE

Crotalus adamanteus

DESCRIPTION

Olive or grayish green, with a chain of large, diamond markings of a darker hue, these are with bright yellow borders about the width of a single scale. The rhombs usually enclose a patch of ground-color; toward the tail they become obscure and finally fuse into cross-bands; the tail is olive, ringed with black. The abdomen is dull yellow.

This snake is not only the largest of all poisonous snakes in Florida, but is also the largest of all venomous snakes to be found anywhere throughout the North American continent. The largest specimen ever recorded and authenticated, measured eight feet, three inches in length. Its diameter was four and a half inches, and the head three and a quarter inches wide. This large specimen was captured in the central part of the state. Specimens of such proportions are extremely rare, in fact it may never be paralleled. The average adult specimen to be found today will average about four feet in length and a six footer is considered very rare.

Compared with the most deadly known species of poisonous snakes of the world, the eastern diamondback rattlesnake ranks second to none. Its huge fangs, and enormous venom glands, represent the maximum degree of deadliness attained by the viperine serpents. The well known bushmaster, *Lachesis mutus,* of tropical South America attains a larger size than this rattlesnake, 9 to 12 feet, and consequently has larger fangs, but careful examination of the fangs of the two species will show that the opening at the tip of the tooth for the ejection of venom is much larger with the *Crotalus.*

The other rival of our big rattlesnake, is the king cobra, *Ophiophagus hanna,* which is found in southeastern Asia and the Philippines. It attains a length of 16 to 18 feet. Its venom acts in a different way from that of the vipers -- immediately attacking the nerve centers.

In structure of the fangs in the diamondback rattlesnake, its near ally the western diamondback rattlesnake, *Crotalus atrox,* and the variety *ruber,* together with the black-tailed rattlesnake, *C. Molossus,* are interesting reptiles, for these snakes have proportionately larger fangs than other venomous snakes of the United States. In this character they relate directly to the South American species of *Lachesis* – the bushmaster – the fer-de-lance – the jararaca and others that have enormous developed, poison conducting teeth.

HABITS & HABITATS

Most deadly of the North American poisonous snakes is both beautiful and a terrible creature, always alert and bold. There is a certain awe-inspiring grandeur about the coil; the glittering black eyes, the slow waving tongue, and the incessant, rasping note of the rattle. It is a rattler that seems to advance upon the intruder rather than retreat. Taking a deep inhalation, the snake inflates the rough scaly body that is usually accompanied with a low rushing sound of air expellation. Shifting the coils to uncover the rattle, this snake is ready to deliver a blow. If outstretched, when surprised, the snake invariably throws the neck into an S-shaped loop with the head drawn back and always within the circle of the body. Various other rattlesnakes of the west, as well as the copperhead and water moccasins will strike from various positions, and will often strike while crawling. The eastern diamondback is capable of accomplishing the same attitudes in striking but mostly persists in its perfectly round and graceful coil, while on the defensive. To observe a large specimen taken unaware and literally fling itself into an offensive position, is to see determination and courage that only exists among few reptiles. Occasionally, though rarely, a diamondback will crawl for cover if disturbed. This sometimes happens when a hiding place is immediately adjacent.

When progressing in normal fashion, this species adopts tactics characteristic of the thick-bodied poisonous snakes generally-slow progress in a perfectly straight line, with head slightly upraised.

Pine swamps, open prairies with clumps of saw palmetto and hammocks are the habitats of these snakes. They often display the habit of hiding under the broad leaves of the palmettoes during the day and only leaving the cool-shaded areas to seek the warmth of the sun or when they venture out for food which begins during the early twilight hours. So closely do the body-colors blend with the vegetation and the affect of sunlight and shadows, that a coiled snake is seen with great difficulty. It will also occupy during hot, sunny days, the deserted burrow of the gopher tortoise.

The favorite food of the eastern diamondback rattlesnake includes the marsh and cottontail rabbits, but other small mammals and birds will be taken. When prey is sought out by the reptile there is no buzzing of the rattle but just like a flash of lightning the bite is delivered. To the human eye the striking jaws seem to have barely touched the victim, but during this blurred moment, several things have happened. The snake struck for the prey with opening jaws, when its head reached the prey its jaws were very wide apart and the fangs raised to such an extent that they were cast directly forward. The fangs pierced the rabbit; the jaws were closed sufficiently to deeply imbed the fangs; a muscle over each poison fang was contracted and a considerable amount of venom was ejected. Sometimes less than one minute passes from the time of the strike to the death of the prey. When the snake is satisfied that the venom performed its use it will uncoil and examine the stricken animal's body until it reaches the head where it seems to make a more detailed investigation touching it with the sensitive forked tongue tips several times. When this procedure is completed the snake seizes the victims nose and the swallowing process begins.

The diamondback rattlesnake gives birth to seven to twelve living young which vary in length from seven to ten inches. These feed readily upon mice at the start, and grow rapidly, fully maturing within two years. All are born with a button at the tip of the tail and venom producing fangs, and with the exception of a more vivid pattern, young specimens are like the parent.

It is believed by many people that the age of the rattlesnake can be figured by the count of the rattle segments. This is entirely wrong, for a new segment is added after each shedding of the old skin. Depending on the environment and the health of the specimen the shedding of the skin and the forming of the new segments can take place as many as four times a year. The use of the rattle to this date is not really known; is it intended as a warning device, a reaction of the nervous system, or is it used in some unknown manner during the mating season.

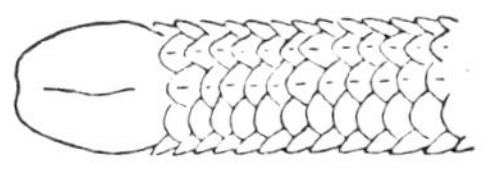

BUTTON OF EMBRYO.

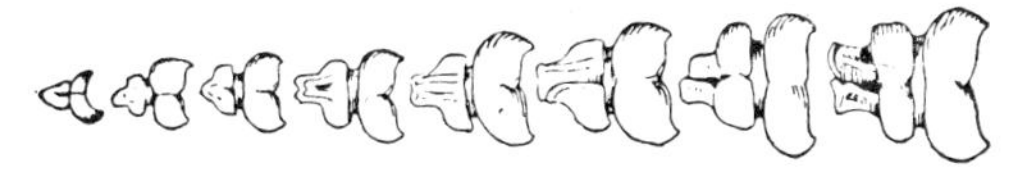

SEPARATED SEGMENTS OF DISJOINTED RATTLE.

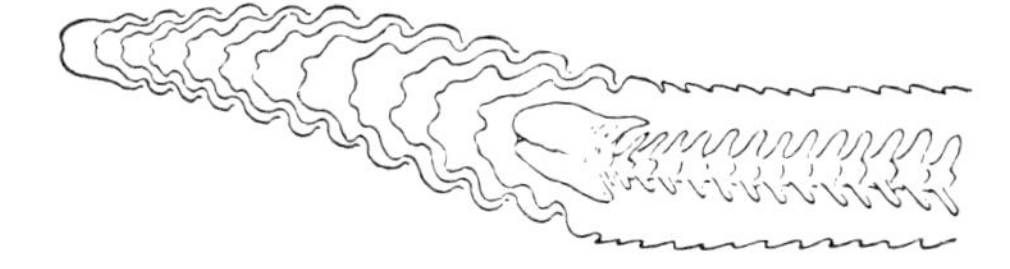

LONGITUDINAL SECTION OF RATTLE.

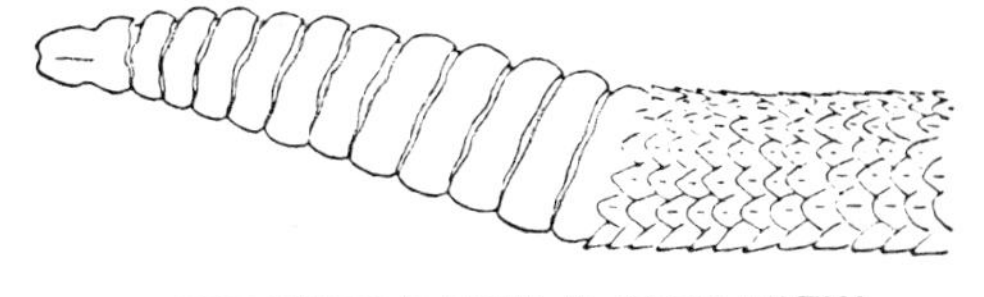

PERFECT RATTLE, SIDE VIEW

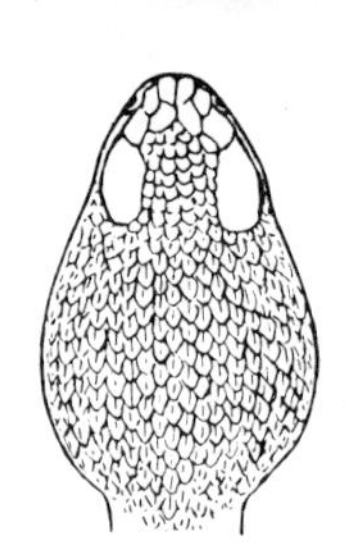

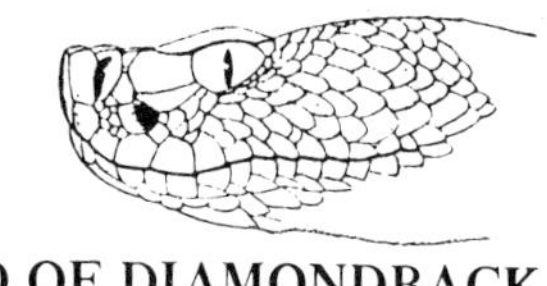

HEAD OF DIAMONDBACK RATTLESNAKE

CANEBRAKE RATTLESNAKE

(Crotalus horridus atricaudatus)

DESCRIPTION

The canebrake rattlesnake is a southern representation of the common timber rattlesnake family, *C. horridus,* that is well distributed throughout the entire midwest and northeastern United States. The canebrake, however, is of an entirely different coloration description despite the likeness in pattern of *C. horridus.*

The canebrake rattlesnake is with a background of grayish-brown to delicate pink, the chevron-like crossbands are jet black; on the center part of the back, for the width of about three scales wide, is a dorsal stripe of rusty red that extends from the nape of the neck to tail. Within the chevron are blotches of ground-color. The chevron pattern continues for about two-thirds of the snake's length then breaks into crossbands of black. The tail is velvety black. The head is with a conspicuous brown band from the eye to beyond the angle of the mouth. Scales are heavily keeled. Length of an average adult specimen is four and one-half feet some specimens have been recorded that were more than six feet.

HABITS AND HABITATS

The canebrake rattlesnake is not really an aggressive snake. In the wilds it prefers escape, though rattling harshly when disturbed, it will usually glide away, sounding its warning note as it goes in graceful form to a place of concealment. If cornered or provoked it will respond bravely, assuming a loose and irregular coil, and striking with such speed that the eye can scarcely follow the movement. It strikes generally one-third, sometimes half its length. While retreating towards shelter it has been known to suddenly turn from a crawling position and, within a matter of one second, draw back the head by contracting the neck into an S-shaped loop and deliver a quick, accurate, venomous bite. However, because of its chosen environment, man very seldom comes in contact with this species.

This snake inhabits the marshy lowlands, wet flatwood areas, river bottoms, and hammocks, in the northern part of the State. Its range extends southward into the Alachua County area.

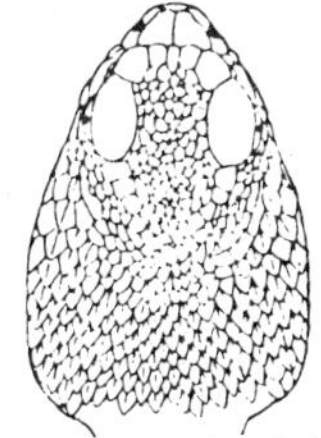

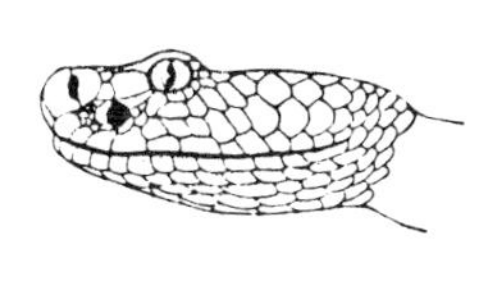

The food of this rattlesnake consists entirely of warm-blooded prey; rabbits, rats, mice, and birds.

Like all rattlesnakes, this species is viviparous – bringing forth seven to twelve young that measure ten to twelve inches long. The young are of the same coloration as the parent and are born with a button which represents the future rattle. All are born with fully developed poison glands and fangs and are capable of delivering a painful bite.

Although considered to be of mild manner this snake should be given the full respect due to all venomous snakes, large or small.

PIGMY RATTLESNAKE

Sistrurus miliarius barbouri

DESCRIPTION

This species and the massasauga *(S. catenatus)* are the only United States rattlesnakes with the crown covered by large plates. However, the massasauga does not inhabit the state of Florida.

A very small rattler, in fact, the smallest of all rattlesnakes, that is stout in form, with a distinct flattened head. The body tapers gradually to a thin tail, which is provided with a minute rattle.

Dark ashy gray, with a series of large, black blotches on the back, these irregularly rounded and separated – on the central portion of the back – by reddish spaces. The reddish broken line, is more prominent on the forward portion of the body. On the sides are several series of black spots, smaller and less distinct than those on the back. The tail is usually reddish. Beneath, this species is white, thickly marbled with black spots and blotches. The large symmetrical head shields, at once, distinguish this small rattler from the young of other species that inhabit their domain.

HEAD OF PIGMY RATTLESNAKE.

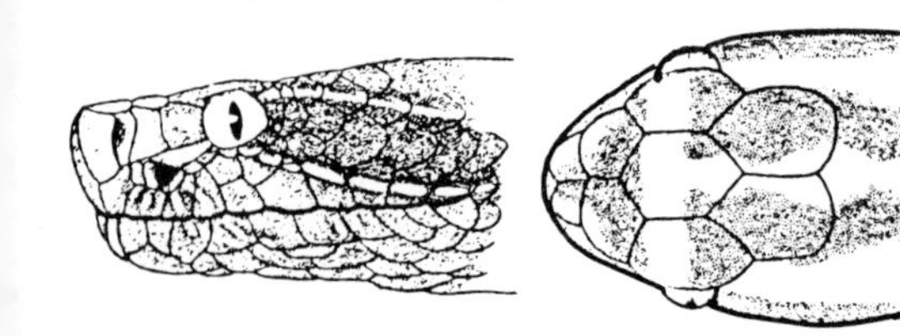

The measurements of an adult specimen would be about seventeen inches in length and the greater diameter, about five-eighths of an inch.

HABITS AND HABITATS

Owing to its diminutive size, this species is considered to be the least formidable of the Florida pit vipers. Despite the deficiency in size of the venom conducting teeth, this little snake is supplied with a powerful venom and should be highly respected. Regardless of the fact that no known deaths from this snake's bite are on record, the bite is very infectious and painful, and remember there is a first time for everything.

So small is the rattle of this species that its whirring can only be distinguished for a distance of a few feet – about eight feet at most with a full-grown specimen, and barely a yard away from a half-grown snake. When annoyed this little snake will throw its body into a fighting coil and sound its tiny rattle, giving vent to its anger by a series of vicious jabs in the direction of the intruder.

This rattlesnake lives in all types of environments, dry bush areas, near lakes, ponds and marshlands. In dry sandy areas they will seek the burrows of rodents and the gopher tortoise to sleep and hide from their enemies during the daytime hours.

Unlike the majority of rattlesnakes, which feed only upon warm-blooded animals, this species is fond of frogs. These it will take with a lightning-like dart, imbedding the fangs deeply and holding the prey until it is dead, when it is swallowed. Frogs that have been bitten and escaped from the snake die within five minutes or so from the effects of the venom which appears to quickly paralyze. This rattlesnake will also eat small rodents, newly hatched small birds, lizards and small snakes.

The pigmy rattlesnake gives birth to living young which number from six to ten per litter. All are born with a minute button and venom conducting mechanism. Young are slightly less than six inches long at birth.

THE RATTLE

By the same method used to determine vibrations per second of a tuning fork, it has been found that the normal rate of vibrations of a rattlesnake's rattle is about 60 per second. It is possible, however, that the rate would be much higher with an irate individual.

The question, "For what purpose does the snake rattle?" is still somewhat unsolved. Possibly it will ever remain so if we continue to look for one single purpose, which may be considered, in the animals welfare as to have brought about development of such a specialized instrument. Researchers, when attempting to explain the advantage for the snake acquiring the rattle, have often failed, because it seemed evident that the rattle, so far from being useful to the snake, in most cases appears to be a disadvantage, which has led to almost total extinction of the rattlesnakes in the cultivated and more densely inhabited districts of the country.

It must not be forgotten, however, that the rattle had evolved long before humankind appeared upon the earth, and that the question of its disadvantages in the struggle against human supremacy could have no influence upon its evolution. The history of evolution is full of similar examples of animals having acquired an advantageous character which, when new predators appeared, was turned against the owner because it could not be undone or modified to suit the new conditions, thus leading directly to its extinction.

The theories of the use of the rattle are numerous, even though we exclude from the discussion the one that it is "natures arrangement to prevent injury to innocent animals and humans." Or is it the other way around? It may be that the rattle is one of the most effective means of self protection, and is useful to the snakes existence as is the growl of a tiger when threatened with danger; truly, awesome sounds of nature!

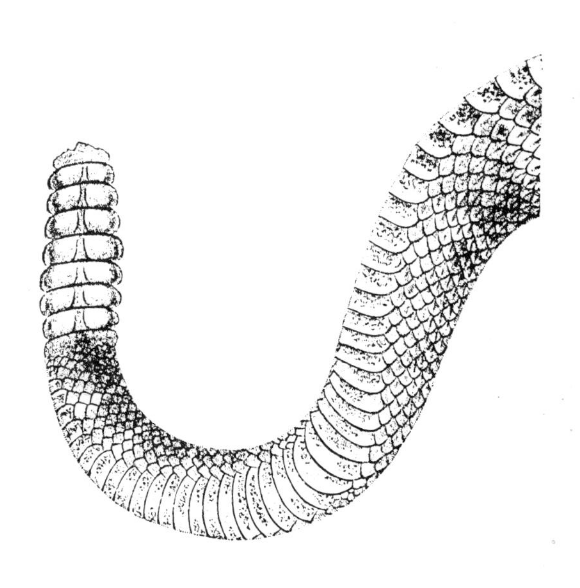

COTTONMOUTH
(Water Moccasin)
Agkistrodon piscivorus

The cottonmouth is a heavy snake in size com parison with other snakes, including the easter diamondback rattlesnake. The head of this poisor ous water snake is very distinct from the neck, an the scales of the body are strongly keeled. On th under portion of the tail, for about two-thirds c its length, the caudal scales are in a single row; th remaining third is with two rows that are arrange much like those of the harmless snakes.

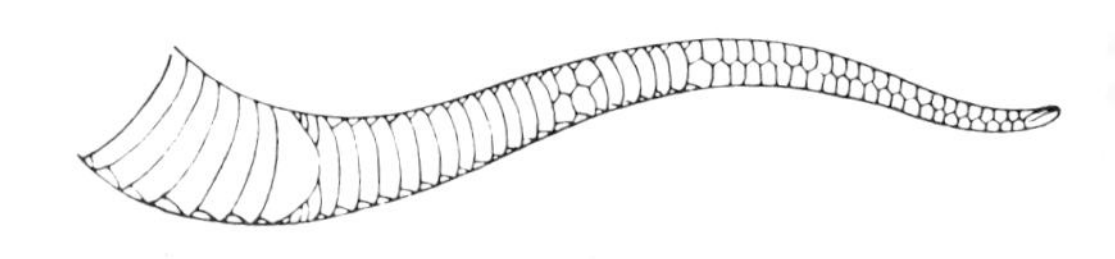

TAIL OF THE COTTONMOUTH
(Underside)

Coloration of the average adult specimen is oliv or brownish above and paler on the sides, on whicl are distinct, wide, blackish bands. These bands en close areas of the ground color and dark blotches.

The upper lip plates are yellow. On the side of th head is a broad dark band from the eye to the angl of the jaw and above this band is a pale yellowish white streak. The top of the head is very dark, usuall black. The chin and lower lip is yellow with thre dark bars on the lip plates, on each side of the mouth A large shield projects over each eye.

Abdomen is yellow, blotched with dark brown o black – more so toward the tail, which is black. Th tail tapers very abruptly from the body.

Half-grown specimens are greenish or brown, witl very distinct bars, while very old specimens ar generally dull olive or black, with little or no trac of the markings.

Although there is a record of this particular snake reaching the length of six feet two inches, the average adult length is about three to four feet. A specimen of this length could measure anywhere from two and one-half inches to three inches in diameter.

HABITS AND HABITATS

The cottonmouth inhabits all of Florida and can be encountered anywhere where there is fresh water; lakes, ponds, marshlands, rivers, cypress swamps and hammocks. They are usually found under or on top of water banks and open flooded areas. Some enjoy resting on a log or heavy brush that projects over the water. When approached it will quickly drop into the water and swim to the bottom or progress across the water to the other side and disappear into the brush. Most accidents are created when people walk, carelessly, near or upon a sleeping specimen.

The water moccasin is very pugnacious when annoyed. The habit of this reptile, when surprised, is to open its mouth, with head drawn back, and disclosing the white mouth parts. This action has been responisble for the name "cotton-mouth" snake. After assuming this attitude, the snake vibrates its tail vigorously.

Unlike the nonpoisonous water snakes that occupy the same habitats, the water moccasin does not confine its diet to cold-blooded prey, but also feeds upon small mammals and birds. Once this snake seizes the prey it retains its hold, with fangs deeply embedded, until its struggles have ceased, then swallowing commences.

The moccasin produces seven to twelve young that almost immediately begin to capture their prey. The pattern at birth is quite different than the parents. They are brilliantly colored. They usually are of a pale reddish-brown, with bands of rich, dark brown. All the bands and markings are narrowly edged with white, making the pattern vivid and striking. It is during this coloration stage that they are often mistaken for a copperhead.

To satisfy an old time argument – there has neve been an authenticated report of a water moccasir while submerged under water, producing a venomou bite upon a human. This sounds very reasonabl because it is impossible for the snake, without som form of solid ground, to accomplish a strike. An with the mouth wide open, a position very necessar for biting, water resistance would force the snak backwards even if he were capable of striking whil under water. However, if a person grabbed and hel the snake under water it would be very likely that bite could be made. This also satisfies an old belie that this snake preys upon game fish such as bas and bluegills. These fish are too fast even for th excellent fast-swimming moccasin. Only the sick an dead fish are taken by this reptile.

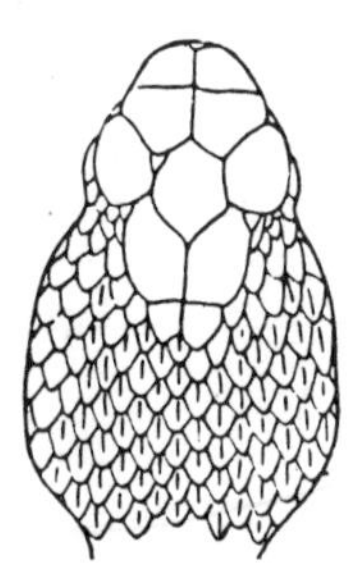

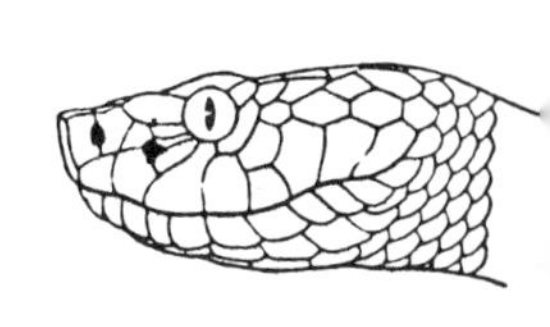

HEAD OF COTTONMOUT

SOUTHERN COPPERHEAD

(Agkistrodon contortrix contortrix)

DESCRIPTION

The southern copperhead is a moderate size snake, and although, of the same genus as the cot-tonmouth, it is a more slender snake. However, lik the cottonmouth the body scales are strongly keeled and the majority of caudal scales are in a single row

Coloration of this particular species is of a pale hue than the other varieties of copperheads of th southwest and southeastern parts of the Unite States. The southern copperhead background color are of pastel shades of hazel brown, buff, or pink ish. There are approximately 16 to 20 large cross-bands of rich, chestnut brown. These bands are narrow on the back and very broad on the sides and, when viewed from above, resemble the outline of hour-glasses. On most specimens several of the posterior dorsal bands are broken, forming inverted-V or Y-shaped blotches on the sides. All of the bands are darker at their borders. In some specimens the bands enclose light patches of the background color. The top of the head is usually with a coppery tinge - hence the popular name. The upper lip is of a shade lighter than the top of the head - the line of intersection between the two colors begins behind the eye and extends to the angle of the mouth. The underparts are pinkish white, with a row of large dark spots on each side of the abdomen.

The southern copperhead is not a large snake. A three footer is considered large, but some rare size specimens have been captured that approached the four foot mark. The greatest diameter of a three foot specimen would be about one and one-half inches.

HABITS AND HABITATS

The copperhead is not a vicious snake but will respond like any other pit viper when its safety is threatened. When disturbed, this snake will usually make an effort to escape. But when escape seems impossible, it will defend itself vigorously. At such times a rapid, vibration of the tail takes place which produces a distinct buzzing sound, if the snake is among dry leaves. This snake like all snakes can strike accurately from any position.

The southern copperhead inhabits both dry and wet forest areas as well as open fields in the extreme northern parts of the State. During the heat of the day they like to seek the cool shadows of tall grass. They can also be encountered under flat rocks, woodpiles and fallen logs.

It has been said by many herpetologists that the venom of the copperhead is not very potent. Don't believe this, I am familiar with tests made upon laboratory animals that have been injected with copperhead venom and in many cases the animal died quicker than those animals that were injected with an equal dosage of water moccasin venom. It must be considered a dangerously poisonous snake that is well able to cause death to man if the bite is well-placed and immediate, competent treatment is not given to the victim.

The feeding habits of the copperhead seem to vary according to the seasons. During the spring and fall it is very fond of frogs, grasping them with lightning-like rapidity and retaining the hold until the prey is dead. During the later spring, it will add young birds to its diet. During the summer months it seems to prefer small rodents. However, like all wildlife, the copperhead is an opportunist and will mix its diet daily if prey of the right size presents itself.

The number of young born to the copperhead varies from six to nine. The young, when born, are about ten inches in length and usually have brilliant sulphur-yellow tails. They are equipped with fangs and their glands are full of venom, and like the parents the young vibrate their tail vigorously when disturbed.

The young as well as the adult specimens are well camouflaged when among dry leaves and one must be very careful when walking through the forest of this snake's domain.

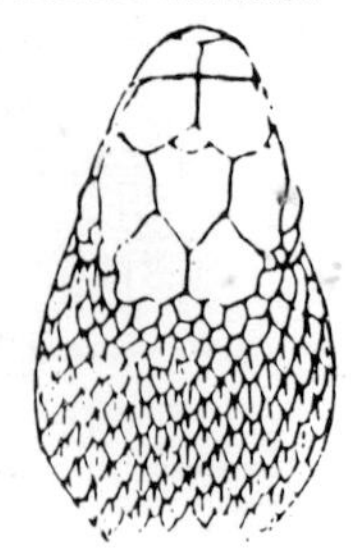

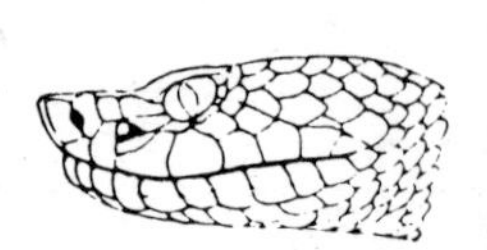

HEAD OF COPPERHEAD.

BANDED WATER SNAKE *Nerodia f. fasciata* - A nonpoisonous water snake that is often mistaken for the cottonmouth. The banded water snake may be encountered in all kinds of watery situations, including fast moving streams and rivers, whereas the cottonmouth prefers sluggish waters.

Photo by Terry L. Vandeventer

SOUTHERN COPPERHEAD

OLD SPECIMEN

Here is an example of a very old cottonmouth that has lost all of its markings. However, totally black individuals, such as this specimen, are considered extremely rare.

KRAITS AND COBRAS

Kraits and cobras are much larger than their near relative, the coral snakes. The king cobra, which reaches a length of 18 feet is the largest of all venomous snakes. Kraits and cobras account for more than 30,000 deaths every year in India. Like the coral snakes, kraits and cobras lay eggs, and the young are venomous at birth.

KRAIT
COBRA

This photo of a canebrake rattlesnake displays the pit and the elliptical eye that is present in all members of the pit viper family.

SCARLET SNAKE *Cemophora coccinea*
A Florida snake that is often mistaken for the coral snake. Pattern above like the scarlet king snake but is with an immaculate, yellowish-white abdomen. Like the latter, the head is sharp, conical and the snout more pinkish than red.

CORAL SNAKE

Micrurus fulvius fulvius

DESCRIPTION

The snout is black and a wide band of yellow crosses the middle of the head. The body pattern consists of broad rings of red and blue-black, separated by narrow rings of yellow that completely encircle the body. On the dorsal sides the red rings usually contain patches of black. The tail contains none of the red rings, being black with broad rings of yellow. The South Florida race, barbouri, although of the same ring pattern, lacks the black pigment in the red rings.

It is, owing to the striking coloration, that the name Harlequin Snake has been given to this species which is slender and seldom attains a length of more than three feet. The head is flat, very blunt, and not distinct from the neck.

HABITS AND HABITATS

The distribution of the coral snake is Statewide. It is of burrowing habit and can accidentally be turned up with a shovel; removing of decayed logs and during the removal of rocks and logs from below the soils surface.

Despite the harmless appearance, the coral snake belongs to a sub-family that contains some of the most deadly known species of snakes. Among its near allies are the cobras, kraits and the Australian tiger snake. All of which are noted for their resemblance to the harmless snakes, and though possessing very small fangs are provided with a venom more powerful than that of the rattlesnakes.

The coral snake bites exactly like their old world relatives, the cobras. Once they have seized the victim, they will advance the fangs into the flesh in a series of chewing motions. This action can produce as many as four to eight separate, venom injected, punctures.

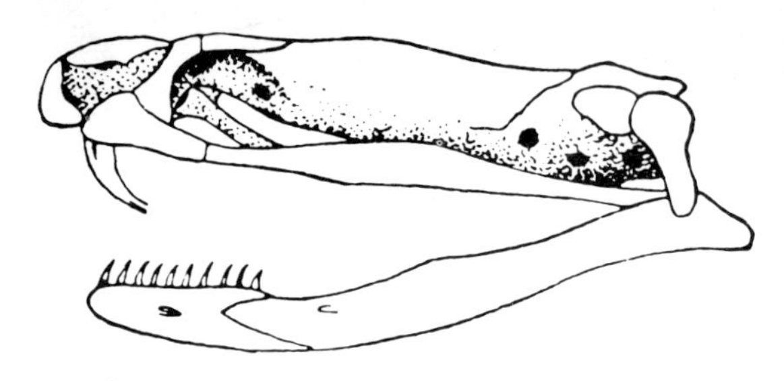

SKULL OF CORAL SNAKE, SIDE VIEW.

The two fangs of the coral snake are rigid and fixed to the anterior part of the upper jaw. Each functional fang is followed by a series of reserve fangs of the same structure as the functional one, but successively smaller. The functional fangs are solidly united to the maxillary bone; being directed backward at a permanent angle of about 45 degrees with the latter. Although small in size the fangs are large enough to be distinguished as being different from the solid teeth of the palate and the lower jaws.

On the front surface of the fangs is a distinct groove. In fact these snakes have been described as possessing "grooved" fangs. This term, however, is misleading and might bring about the idea that the structure of the fangs is the same as those possessed by the rear-fanged snake group, *Opisthoglyph.* Although the face of the fang is deeply furrowed, the venom conducting teeth contain a canal for the flow of poison and open in a small orifice at the tip, in the same fashion as a hypodermic needle.

The venom of the coral snake is a neurotoxin which means that the venom is composed mostly of poisons that affect the nervous system causing paralysis and sometimes blindness. Its venom is considered, by many experts, as being the most dangerous of all Florida snakes.

It is believed by the misinformed that this beautiful reptile is docile in disposition, seldom attemtping to bite. This belief is dangerously misleading and has been proved time and again that it is not a reptile to trust. **(continued on page 45)**

EASTERN DIAMONDBACK RATTLESNAKE

This is the threatening attitude of the diamondbac
rattlesnake. Notice the great number of segmen
that form its rattle. The true age of a rattlesnak
cannot be accurately told by the count of segment

**THREATENING ATTITUDE
OF THE CANEBRAKE RATTLESNAKE**

CANEBRAKE RATTLESNAKE

PIGMY RATTLESNAKE

This pigmy rattler is testing the air with its forked tongue, either for particles of odor that may direct the snake to its next meal, or for vibrations that will warn it that a predator is closeby.

COTTONMOUTH
(Water Moccasin)

COTTONMOUTH

This is the traditional threatening attitude of the cottonmouth water snake. Some specimens may not display their cotton-white mouth parts, but never-the-less, when disturbed, all individuals are ready to bite.

YOUNG COTTONMOUTH

The pattern of this young cottonmouth is very bright, however, as the snake reaches maturity much of the pattern will become obscure and eventually, with age, may be lost entirely.

All snakes like to rest in an animal burrow, where they are usually safe from predators and can remain cool during hot, sunny days. This photo shows a diamondback rattlesnake resting in a burrow, made by a gopher tortoise.

Photo by Terry L. Vandeventer

Be very careful where you put your hands or feet EVEN IN YOUR OWN BACK YARD.

REMEMBER that the nose of a coral snake is always **BLACK.**

SCARLET KING SNAKE *Lampropeltis triangulum elapsoides* — This snake is often confused with the deadly coral snake that inhabits the same areas. Pattern, red and yellow rings broad; the yellow bordered with rings of black, more or less encircling the body. Snout red and pointed.

(continued from page 32)

It is true that a snake of this kind may be handled without accident, as its actions in biting are quite different from many snakes. Nevertheless, the great majority of bites inflicted upon humans have been mostly accomplished during this practice.

After heavy showers and at night the coral snake ventures from its resting place to forage for food which consists chiefly of snakes and lizards. It is especially fond of the young five-lined skink, often called the blue-tailed skink, which creeps under the loose bark of decaying logs and fallen timber for the night. When feeding the coral snake displays a ferocity of motion quite contrary to its general sluggish disposition. If the prey be a snake, it is quickly seized by the neck or body and the fangs advance towards the victims head with a series of chewing motions. This action injects considerable venom and prepares the struggling prey for the swallowing process. The coral snake has been observed swallowing a snake that was but a few inches shorter and of greater girth than the feeding reptile. When this is done, the coral snake becomes rigid and is unable to coil itself fully for at least two or three days. In fact it will find it almost impossible to crawl to its shelter to sleep off the greatest part of the digestion period.

The coral snake is oviparous. Its eggs are white and greatly elongated. They are deposited in decaying bark or loose, damp soil. Incubation period is about 90 days. The young, when hatched, are about seven inches in length and approximately one-eighth of an inch in diameter. They become very active within a matter of minutes after birth and are vicious biters that are able to induce a lethal dose of venom into small prey.

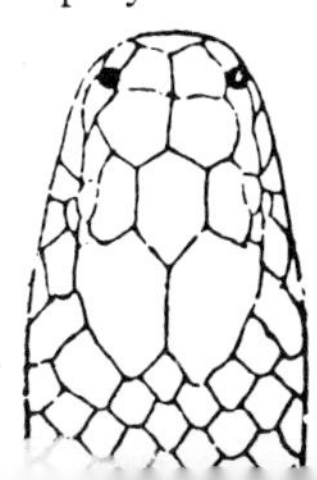

HEAD OF CORAL SNAKE.

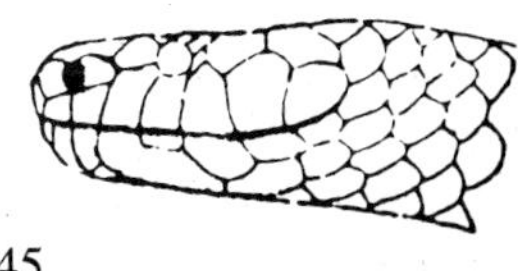

CROWNED SNAKES
(Rear-fanged Snakes)

Crowned snakes are small, reddish brown or tan snakes with a black cap. They are poisonous, rear-fanged snakes that are not dangerous to humans. There are five species that inhabit the state of Florida, but because of their close resemblance to one another, an overall description will be given on the most common species; the remaining four species will be described with illustrations of the head markings that separate them. Their habits, habitats, and range will also be included.

CENTRAL FLORIDA CROWNED SNAKE
Tantilla relicta neilli

A smooth scaled snake with an average length of 8 inches. Body color is ususally pale reddish brown to tan. The head is somewhat flat, and not distinct from the neck - it is dark, almost black, with an obscure yellow to white crossband, that is followed by a narrow, black band on the nape of the neck. Underparts are white, but may have a tinge of pink or yellow. There are two points of identification on the crowned snakes that distinguishes them from the brown snakes and the ringneck snakes: 1. All crowned snakes have 15 rows of scales throughout the entire length of their body. 2. All are without a loreal scale (a small scale between the preocular scales), a scale that is present on the brown and ringneck snakes.

Crowned snakes prefer loose soil or leaf mold situations, in and around forest areas. They are secretive snakes that can be found under leaf mold, rocks, loose bark, and all kinds of debris. These small snakes show no desire to escape and never attempt to bite while being handled. Food consists of centipedes, and most soft bodied insects and their larvae that live below the surface of the ground, under rocks, bark, and leaf mold.

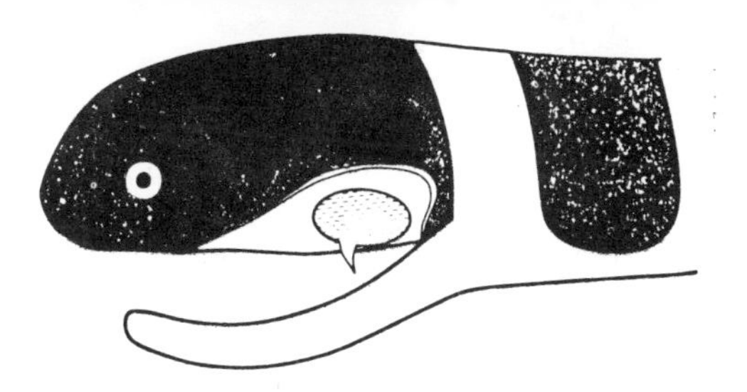

Head of crowned snake showing rear fang and venom gland.

Crowned snakes are with two, very tiny, grooved fangs, that are positioned in the rear part of the upper jaw. Each fang makes contact with a venom gland that is situated immediately above the fangs. Venom is highly toxic and very sufficient for subduing their prey, but it is practically harmless to humans. However, if bitten, the person should be treated by a physician.

The heaviest concentration of recordings of the central Florida species have been in the eastern half of the peninsula; Suwannee County to the north, and south to Hillsborough County, from the drier areas of the Gulf coastline, eastward to the mid-state counties.

FLORIDA CROWNED SNAKE
Tantilla relicta neilli

The black cap of this species is separated from the nape band by a much less distinct, light, line than that of the southeastern species. This snake is most often found under rocks and debris of soft soil fields and open woodlands of the peninsula.

SOUTHEASTERN CROWNED SNAKE

Tantilla coronata

Found in the western panhandle from Escambia County, eastward to Calhoun County. Same habitat as the central Florida species.

PENINSULA CROWNED SNAKE

Tantilla relicta relicta

Scrub areas of central Florida, from Marion and Flagler counties, south to Highland County, and northeast to Brevard County. Reports, however, show that this snake has two small populations in Charlotte and Sarasota counties to the southeast, and Seahorse Key in Levy County, to the northeast.

COASTAL DUNES CROWNED SNAKE
Tantilla relicta pamlica

From Brevard County south to West Palm County. This species inhabits the semi-arid grass and bush covered dunes, and scrub areas of the coastal regions.

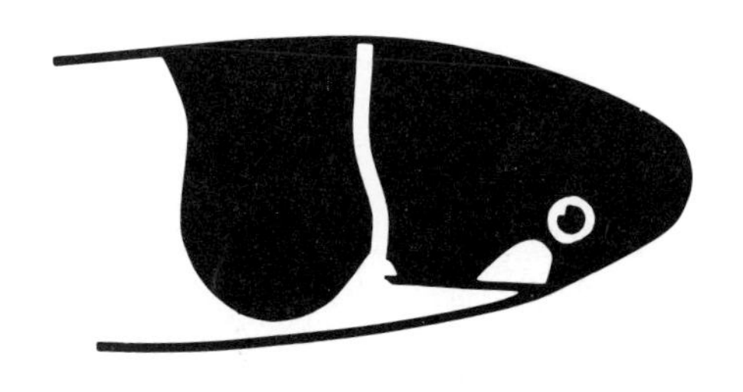

RIM ROCK CROWNED SNAKE
Tantilla oolitica

From the extreme southeastern part of Florida; Dade County and Key Largo. As its name implies, it inhabits rim rock or oolitic limestone areas that are with plenty of plant cover.

CAUTION

Florida is a State that abounds in beautiful lakes, rivers, woodlands, fields and swamplands. Every year millions of residents as well as visitors visit these wonderful places as part of their vacation, work, play or to enjoy their favorite sports; hunting, fishing and golf. Whatever your pleasures might be, you must remember that there is always a possible chance of coming in contact with a poisonous snake.

Most snake bites are accidental and as we know accidents are most often caused by carelessness, lack of knowledge or just plain ignorance.

When hiking in the fields or woodlands always step up on any obstacle and look around before proceeding on. Snakes like the cool ground that lies in the shadows of fallen logs, piles of rock and debris. Be careful of all hollows under trees and rocks as well as burrows, like those of the gopher tortoise and the burrowing owl. All snakes like to crawl into an abandoned hollow or hole to get out of the hot sun and away from their enemies.

When fishing in remote fresh or brackish water, lakes, rivers, canals and swamplands watch where you step – frequently the deadly moccasin will be basking in the sun or in the grassy areas along the waters edge. When in water moccasin areas never enter the water from a high bank – always try to enter at a shallow shoreline or beach area. Also what might look like a good crayfish, catfish or turtle hole, under the water's bank could be a hide-a-way for the deadly cottonmouth.

When camping always remove possible firewood off its resting place with a stick before picking it up with your hands – a snake might be in or under it. If using a tent make sure it has a good one-piece flooring and a good closing entrance, zipper preferred, and by all means pick a clear, open area to pitch it. When using a camper, truck or house trailer, pick a good open area and inspect the ground under it occasionally for any snakes that might have found shelter from the hot sun during the day's heat or during the cold night.

When hunting in diamondback rattlesnake country a good pair of snake-bite proof boots should be worn. Old boots that have become softened around the ankles, from long periods of use, should never be trusted. Large specimens of diamondbacks and water moccasins have tremendous striking power and are able to sink their sharp, long fangs through the leather and into the person's flesh. Most bites from all snakes are most often inflicted in the ankle area.

Never attempt to kill a poisonous snake with a stick, this usually places an inexperienced person within striking range of the reptile without realizing the capabilities of the length of an irate snake's strike. I witnessed a person trying to accomplish this act with a four foot length of heavy branch. The diamondback that he tried to kill struck out and missed his hand by a few inches.

Last but not least – always give a poisonous or unidentified snake plenty of room and the right of way. This is a positive precautionary measure to prevent unnecessary accidents.

SNAKEBITE TREATMENT

On the average, about 200 snakebite cases are treated every year in the Sunshine State. Not all of these bites are by poisonous snakes – those that were by poisonous snakes, not all were with fatal results. Some victims of non-poisonous snakebites suffer infection, extreme shock, or both. While those who suffered from poisonous snakebites experienced local swelling, usually accompanied by extreme pain prior to recovery or death. Some who recovered suffered amputation of a hand, arm or leg. Most, however, recovered with no side affects at all. In fact about 98% of venomous snakebite victims survive.

NONPOISONOUS SNAKE BITE.

The bite from a non-poisonous snake can be recognized by the pattern of scratches or punctures caused by the upper and lower rows of teeth. A non-poisonous snake bite should never go untreated, any good antiseptic can be used to help prevent infection.

POISONOUS SNAKE BITE.

In the bite of a poisonous snake there will be one or two distinct punctures present. If venom has been injected, unmistakable symptoms will quickly appear. Usually there is a burning sensation with some pain in the vicinity of the bite followed in minutes by a swelling of the bite area, directly proportional to the amount of venom injected; discoloration of the skin, usually followed by nausea and vomiting. However, some pit vipers have been known to produce temporary blindness or paralysis, more so in the case of a water moccasin bite, than from the bite of a rattlesnake or copperhead.

It is not for me to recommend any one special treatment of cases of snake-bite poisoning; but I will state what would be considered rational treatments, in view of our present knowledge of the subject. When anyone enters this field, they will find it a field in which controversy and argument among medical authorities prevail. However, the following methods are accepted by many of the medical profession as well as herpetologists all over the world.

Anti-venin is the most effective treatment for a poisonous snake bite. However, if at all possible, the injection should be administered by a competent person or by a physician. Anti-venin is a foreign protein to which the individual may suffer considerable adverse or even fatal reactions. Remember, it is never too late to administer proper treatment. Lives have been saved when the victim was close to dying. The real danger is in the delay of proper treatment.

In the event of a snake bite try to identify the reptile. If you are not familiar with snakes it should be killed, with the least exertion possible, and kept for proper identification by a responsible person. If this can not be done, attempt to describe the snake by writing out the best description possible. Trying to remember the pattern after the bite will remain secondary in your mind and can easily be forgotten due to the emergency at hand.

Make sure that some symptoms of envenomation have actually occurred before starting any field treatment. Incisions about one-quarter of an inch long should be made across each puncture. The incisions should never be more than skin deep, not into the muscle. Suction by means of a first-aid device or by mouth should be started immediately. A tourniquet can be applied approximately four to six inches above the wound and should never be made so tight as to cut off the blood circulation completely, this can prove dangerous. The tourniquet should be released every three to five minutes for a one minute period, then again tightened. The suction should not be continued for more than 15 minutes, it has been proven that after ten minutes very little, if any, venom can be withdrawn. All of this treatment should be done, if at all possible, during your movement to the nearest physician or hospital. However, there are many case histories of complete recoveries of envenomated persons that made no use of the incision, suction and tourniquet method. Their recovery was rapid by immobilizing the part of the body bitten and getting to a hospital as quickly as possible. Most of all do not panic into shock, after all, very few poisonous snake bites prove to be fatal.

If antivenin is available, in the field, be sure to read the directions carefully before making any injections. If the person bitten is known to be allergic to horses, injections should not be attempted, instead, make for the nearest hospital for proper treatment. Antivenin contains horse blood. However, if instructions are followed and injections made, the patient should still be taken to the nearest hospital for further treatment.

In the event of a coral snake bite, symptoms may not be noticable for approximately one to two hours following the bite. In some cases the span has been greater. The first sign of envenomation is a soft swelling at the puncture site. Dangerously advanced symptoms of this deadly snake's bite are blurring of vision, drooping of the eyelids, unsteady gait, thick tongue feeling, slurring of speech, tingling sensations, nausea and vomiting, temporary blindness and paralysis.

In the State of Florida the overwhelming majority of persons bitten by snakes are within an hour or two at the most, from a hospital emergency room where treatment can be administered under the supervision of a physician.

DISTRIBUTION

EASTERN DIAMONDBACK RATTLESNAKE
COTTONMOUTH
CORAL SNAKE
Statewide including some of the Keys

PIGMY RATTLESNAKE
Statewide — rare in the Keys

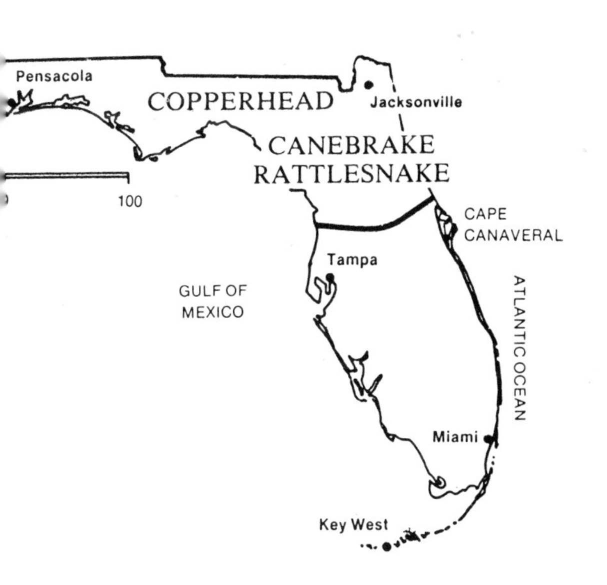

RECORD LENGTHS OF POISONOUS SNAKES

SPECIES	LENGTH	METRIC
CORAL SNAKE, Eastern		
(Micrurus f. fulvis)	47½"	120.7 cm
CROWNED SNAKES		
Peninsula		
(Tantilla r. relicta)	10¼	26.04 cm
Central Florida *(T. r. neilli)*	9½	24.1 cm
Coastal Dunes *(T. r. pamlica)*	9	22.86 cm
Southeastern *(T. coronata)*	13	33.00 cm
Rim Rock *(T. oolitica)*	11½	29.2 cm
PIT VIPERS		
Cottonmouth		
(Agkistrodon piscivorous)	74½	189.2 cm
Copperhead, Southern		
(Agkistrodeon c. contortrix)	52	132.1 cm
Pigmy Rattlesnake, Dusky		
(Sistrurus miliarius barbouri)	25⅛	63.8 cm
Canebrake Rattlesnake		
(Crotalus horridus atricaudatus)	74	189.2 cm
Eastern Diamondback Rattlesnake		
(Crotalus adamanteus)	96	243.8 cm